Now is the Time for Modern Lift Stations

Why your community needs to upgrade to innovative systems

Jon Dunham

© 2015 by *Jon Dunham*

All rights reserved.

All Rights Reserved. No part of this publication may be reproduced in any form or by any means, including scanning, photocopying, or otherwise without prior written permission of the copyright holder.

Disclaimer and Terms of Use: The Author and Publisher has strived to be as accurate and complete as possible in the creation of this book, notwithstanding the fact that he does not warrant or represent at any time that the contents within are accurate due to the rapidly changing nature of the Internet and the industry. While all attempts have been made to verify information provided in this publication, the Author and Publisher assumes no responsibility for errors, omissions, or contrary interpretation of the subject matter herein. Any perceived slights of specific persons, peoples, or organizations are unintentional. In practical advice books, like anything else in life, there are no guarantees of income made. Readers are cautioned to reply on their own judgment about their individual circumstances to act accordingly. This book is not intended for use as a source of legal, business, accounting or financial advice. All readers are advised to seek services of competent professionals in legal, business, accounting, and finance field.

First Printing, 2015

ISBN-13: 978-1517483760

ISBN-10: 151748376X

Printed in the United States of America

Dedication

To those persons who have an interest in providing the best in handling of wastewater and modern services to their home city, the persons can be housewives, homeowners, mayors, city employees, sewer workers, or any other person who has an interest in reducing costs that are ultimately born by the ratepayers and/or taxpayers.

This publication is to offer a quick study of what is now available to replace the systems that have been used in this country since the 1930's. The world is changing and the methods to handle wastewater have also changed. There is no reason to continue to use the same old submersible pumping systems in wet wells that have not had a basic design changed in over 50 years.

Wastewater has changed from handling just the human waste and easily disintegrating paper products to now being required to process "flushables" and many other disposable products through the waste system.

The new and modern lift stations handle the flushables and at the same time, save tremendous costs of operation. It's time to modernize.

Now is the time for modern Lift Stations

Why your community needs to upgrade to innovative systems

Table of Contents

Introduction 9
Chapter 1 Manage; Not just Monitor 15
Chapter 2 End Wasted Costs.. 17
Chapter 3 Stop Unclogging Pumps 19
Chapter 4 Stop Cleaning Wet Wells 21
Chapter 5 Easy Grit Removal 23
Chapter 6 No More Float Switches 25
Chapter 7 Change Impellers in 15 Minutes..................... 27
Chapter 8 End the Corrosion 29
Chapter 9 End the Odors 31
Chapter 10 Still Using Since the 1930's 33
Chapter 11 End FOG Issues... 35
Chapter 12 End Trash Hauling 37
Chapter 13 Did You Get the Call?.................................... 39
Chapter 14 A Converted Wet Well 43
Chapter 15 Start & Stop 150 Times an Hour................... 45
Chapter 16 Run Dry for 150 Hours 47
Chapter 17 Made of Stainless Steel 49
Chapter 18 Manage Remotely 51
Chapter 19 Managing Remotely...................................... 53
Chapter 20 The Sensor System 55
Chapter 21 Assembled in the Heartland 57
Chapter 22 Compare with what you have....................... 59
Chapter 23 Your Cost of Operation................................. 61
Chapter 24 What you can gain 63
Conclusion 65
About the Author.. 67

Introduction

As announced via the AWWA, for the fourth year, "the top priority is the aging of the infrastructure or our water systems." In addition to that, "the increased age of the current operating employees in water systems is growing more worrisome as young replacements are not coming up."

"Costs to upgrade equipment and machinery are major considerations as funding is not readily available."

By spending a few minutes here, you will learn of an innovative, sustainable wastewater pumping system that will address the issues mentioned above. When your wastewater lift stations are in need of an upgrade, or you're growing tired of sending a crew out again and again to de-rag or clean the trash out of a wet well, do you ever dream of not having to do this again?

Do you have to answer to the neighborhood about what you are going to do about the smell? What are you going to do about the corrosion in the wet wells? Why do you spend the money on a "vac-truck" service every time the maintenance schedule says you have to do this? Would your people be happier if they weren't on call so often? What would it be like to never have to deal with a wet well again? This can, in today's world, be your reality.

Given the likely impending installation of many technologically advanced lift stations in the United States, municipalities and industrial users will recognize that their savings of manpower costs must gain greater importance.

This is done by implementing a pumping system that no longer requires having to remove and replace clogged pumps by installing pumps that are self-cleaning.

By "self-cleaning" I mean cleaned without human intervention, done remotely from anywhere. Additional savings will accrue as the corrosion of lift station equipment and the station itself will cease. The use of bar screens and rakes, trash baskets, grinders or comminutors, all requiring regular cleaning and maintenance at the wet well is no longer needed.

Employees can now do more productive work and not the "dirty" jobs as in the past. Now, by using modern pumping systems, your operation is even more attractive to prospective new employees.

The daily issue with "flushables" causing problems with lift station pumps will no longer need personnel for on-site, regular attention. There are no more float switches that attract grease and require cleaning.

The neighborhoods will experience the added relief and benefit of no longer having the hydrogen sulfide gas odors. Odors are due to the age old required local retention of raw sewage in a wet well as has been the practice through using submersible pumps for the last 60 years. It is now time to consider a modern system and make a change.

The need for a wet well is also eliminated and it is converted to a clean, dry pumping system work area that is now safe for employees to enter if needed. A separate valve vault is not needed and the footprint of the station is smaller. The new, modern lift station is remotely managed, not just monitored, as some submersible systems currently report. The management can be done via a smart phone, a tablet, or a desktop computer from anywhere. The operator has the ability to visually "see" the performance of the pumps, make any adjustments if desired, run tests, and even order printed reports to be delivered to management via remote printer from his office, the cab of his truck, his kitchen table or while watching the game on a weekend.

The typical daily or routine lift station "site visits" are no longer needed. This modern system provides effective utilization of personnel and supports optimal usage of power in its operation. The proven electronic control techniques employed use soft starts and stops that eliminate water hammer shocks to valves and older force mains. Not like a full speed start up and an abrupt shut off as is typical of submersible installations. The modern, up to date pumps only run at the speeds needed and no longer is a wet well pump down type of application used. No large surge of wastewater will now affect the biomass of the treatment plant making the operator's job easier and less costly. The effective utilization of the voltage control techniques will maintain power quality and ensure that the pumping equipment continues to operate efficiently. This modern system will contribute to maintaining an efficient lift station with high quality and reliability even under future scenarios involving varying peak loads as it is designed with and includes 100% back-up pumping capacity.

Increased demand for clean water adds additional burdens on the aging systems in place in the United States. In the management of wastewater, typical wet well lift stations have surpassed their useful lives and continue to demand the costs of maintaining and operating at or over their designed capacity. While the odorous wet well design of many years ago is still in use, lifestyles and sanitary products have made great changes and additions. We now experience an era of "flushables' that use the waste system as a disposal method. These lifestyle modifications are not likely to change even if the construct of the items flushed is altered. Wet well lift station use and maintenance costs here in the United States will continue to climb. As long as wet well use is continued, the maintenance of it will be required. Why not eliminate this ongoing cost?

It may be of interest to note that many of the European cities have turned their water and wastewater operations over to private industry. When you realize that these private contractors are in the water and wastewater business to make a profit, and they do, does it make you wonder how?

The answer is easy…They have adopted the modern, up to date pumping systems as described above. They no longer use the old costly submersible wet well system. Obviously, they would not do this if it didn't make them money! To you, "profit" may mean using less tax dollars or providing better service for your ratepayers.

The actual case studies that are available will easily give one comfort in knowing that many others have profited in their implementation of modern equipment.

Reducing the cost of ownership is one of the keys to being a successful manager. Happier employees doing less of the "dirty" work is another. Working for an operation that employs modern technology also is recognized for its leadership and will continue to attract potential new employees.

I have spent time visiting cities and meeting with their operations, engineering and management personnel. I have had "Lunch and Learn" programs with engineering groups. I have met with state officials. I have met with the federal agencies and presented the system. I have contacted all of the agencies that I can find. I feel that it's important to my country, my planet, and you, to know that there is a better way to pump sewage.

It's now time for answers and solutions. I can provide both. I will demonstrate how you can save costs and at the same time use this innovation to benefit your ratepayers or any others who have an interest in how you profit.

You will now be able to respond affirmatively to the agencies that continually ask for "innovation and sustainability." You will have done your job.

Chapter 1

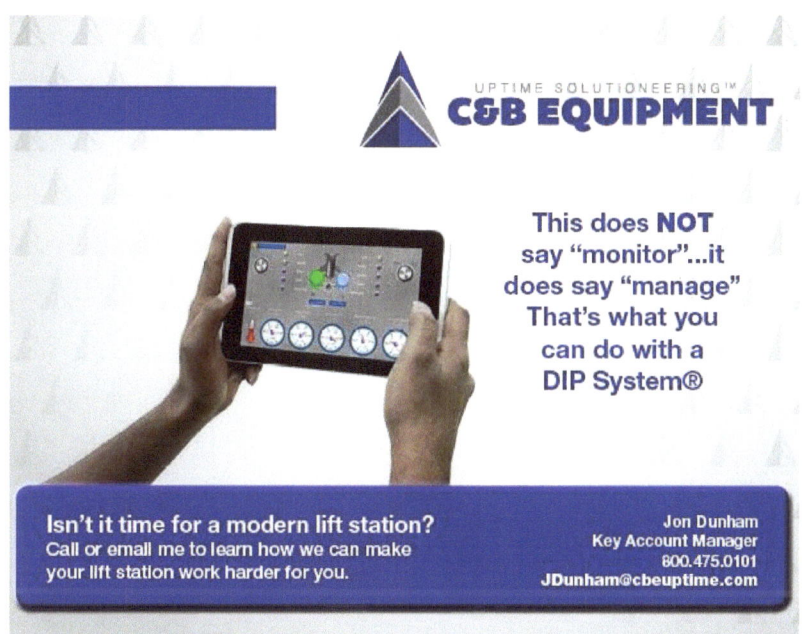

Current wastewater systems typically only can be monitored for their operations. They are not capable of remote management.
The DIP System® lift station is remotely managed, not just monitored, as submersible systems currently report. The management can be done via a tablet or a desktop computer from anywhere. The operator has the ability to visually "see" the performance of the pumps, make any adjustments if desired, run tests, and even order printed reports to be delivered to management via remote printer from his office, the cab of his truck, or his kitchen table on a weekend. The typical daily or routine lift station "site visits" are no longer needed.

…And yes, you can even manage remotely with your smart phone. Just like using a tablet or a desktop computer.

Consider the hours saved in all of the trips to the lift stations. Consider the cost savings of just using the remote management features. Why would you not want to modernize your wastewater pumping?

Chapter 2

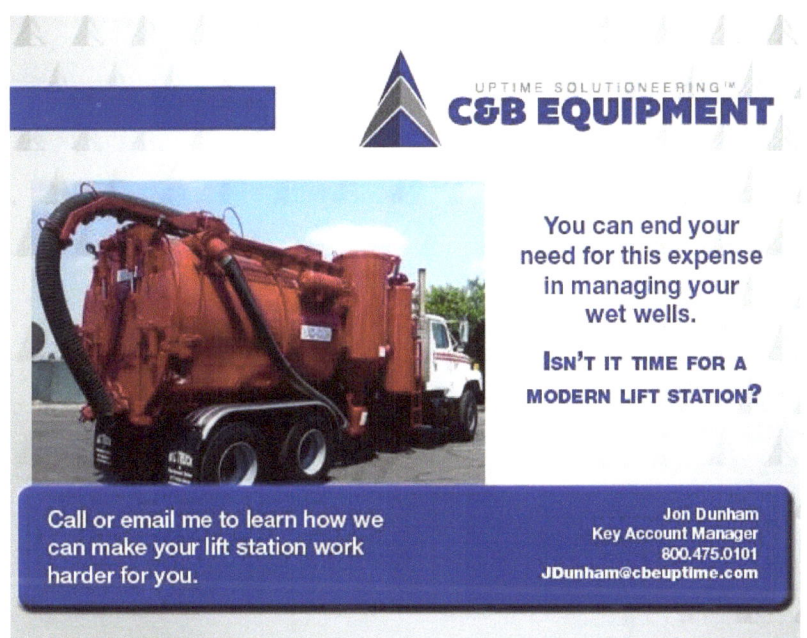

The need to clean a wet well varies from once a year to multiple times if there is a grit problem or other infiltration of fine debris. Use of the vacuum trucks is expensive and never ends as long as you have and use wet wells. These costs can be eliminated with a modern lift station that uses the DIP System®. You will never have to clean a wet well again. How much can you save by modernizing?

Chapter 3

Europeans have used "wipes" for many years. The same pump clogging issues that are being experienced in the United States today were addressed and resolved years ago by a French innovator, Stephan Dumonceaux. His company, S.I.D.E. Industries, did something about it. In fact, the first DIP System® was patented, installed and proven successful in 2003 and since then there are over 1,300 of these systems in use in municipalities (as well as Disneyland Paris) throughout France with many more to come.

Other municipalities in other parts of the world are also now relieved of the constant cleaning and unclogging of their lift station pumps.

The DIP System® pumps actually clean themselves, automatically, without human intervention.

Why would you continue to spend the money to pull and clean pumps when you could end this practice? What other, cleaner jobs, would your personnel be doing if they didn't have to unclog pumps?

How much can your municipality save by modernizing?

Chapter 4

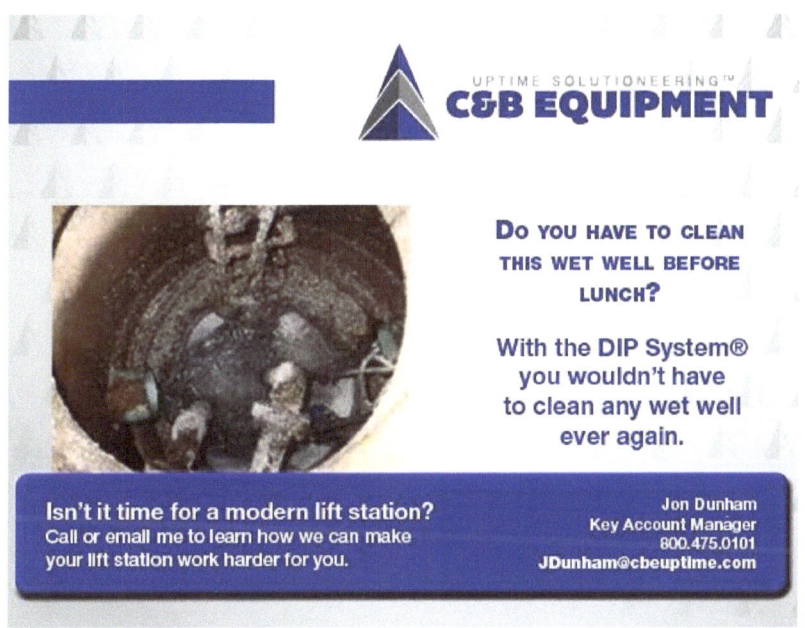

The United States has been using wet wells with submersible pumps since the early 1930's. This same "technology" is still used as part of our aging wastewater systems today. What is a wet well? It is a typically round, concrete cylinder, 8 feet or more in diameter, and about 25 feet deep and found in neighborhoods and in city lots. It will usually have two pumps (duplex) that are submerged to the bottom of the well. Sewer pipes from surrounding homes and businesses drain their raw sewage, after flushing, by gravity into the wet well. The wet well is used to collect and retain the raw sewage until a predetermined level is reached and it is then pumped down and sent on to the treatment plant. The wet well can now fill again.

The neighborhoods will experience the added relief of no longer having the hydrogen sulfide gas odors due to the required local retention of sewage in a wet well as has been the practice through using submersible pumps for the last 60 years. The need for a wet well is also eliminated and it is converted to a clean, dry pumping system work area that is now safe for employees.

Why would you want to continue to retain raw sewage in the neighborhoods? Why would you want to continue to deal with the odors? Why would you want to have the need to clean a wet well? This can end when you modernize with a DIP System®.

Chapter 5

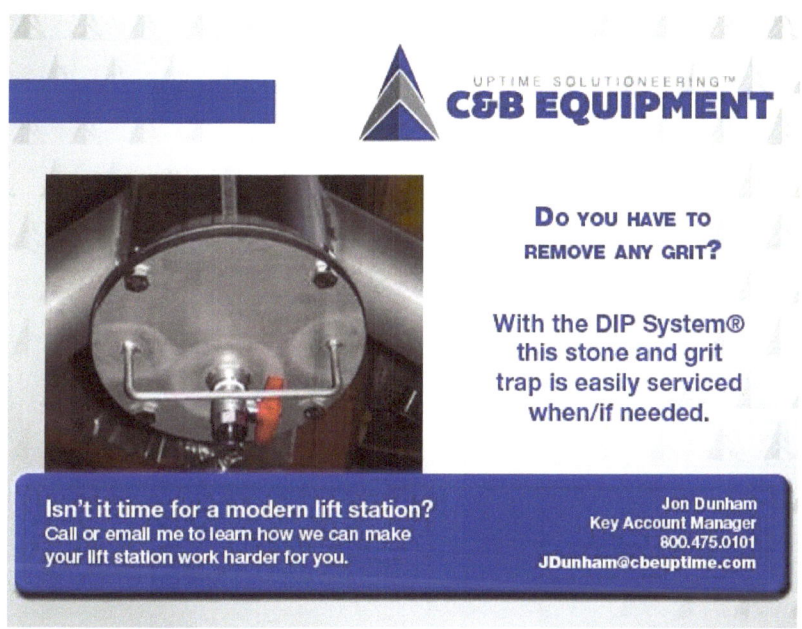

The DIP System® bodies' suction profiles are specially designed to take advantage of the flow speed from the gravity-driven inlet. The inlet body also serves as a grit and stone trap with inspection port and draining valve. The interior surface of the body is very smooth to improve efficiency and doesn't have any areas where matter in suspension might collect.

With a built-in grit trap, removal is simple and easy. Why have all of the additional costs to remove grit?

Chapter 6

Are you cleaning your float switches at your lift stations? Is the grease a problem? How much do you spend on this action in your pumping operation? Did you know that float switches are not even needed in a DIP System®? Why would you want to continue to spend this additional cost for as long as you have a wet well? Doesn't it make sense to modernize?

Chapter 7

Change an impeller in 15 minutes? Yes you can.

DIPCut® is a patented impeller that becomes a "Shredder" when it changes direction of rotation. Coupled to the variable frequency drive, this impeller changes its direction of rotation automatically when needed in order to cut snarled long fibrous materials and rags, and then removes them.

 DIPCut® combines the advantages of the conical Vortex impeller when pumping sand, gases or big solid wastes and the shredding function cutting long fibrous materials into shreds. The switch between the 2 functions is managed automatically by the drive controller based on the torque control and monitored via the OmniDIP® Box connected into the command.

Moreover the self-monitoring system OmniDIP® allows the remote follow up and analyses of the functioning of the DIPCUT®.

Impeller changes are not regularly needed, but why would you want another type of pump…With the old common submersible pumps, you have to pull it up, disconnect wiring, take it to the shop or have special tools at the site to change the impeller.

With the DIP System® you only need a few minutes and common hand tools. How much can you save?

Chapter 8

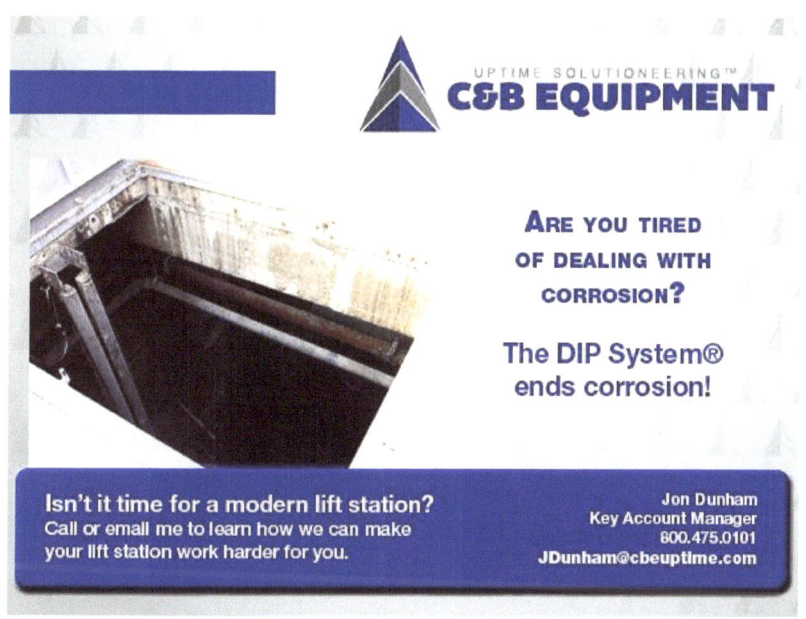

To renovate an existing wet well lift station the DIP System® adapts to any type of currently available pipework so precise positioning of input/output pipes is no longer required. The discharge head can be positioned at any angle through 360°. By eliminating the wet well, the dry tank becomes an equipment room which can be fitted with lighting, a ladder and other accessories which enable maintenance personnel to carry out their work in safety.

The former wet well lift station now becomes a straightforward inspection chamber without human danger as there is no emission of dangerous gases, odors or accumulation of solid matter.

The equipment is rustproof, and is thus more resistant and more durable. The absence of a collection and retention tank eliminates costly cleaning operations of those traditional units. The DIP System® enables flow or discharge capacities to be increased in a pumping station which has insufficient power, without changing the civil engineering.

With no longer having to deal with corrosion, what can you save? No more replacement of disintegrated components. No replacement of electrical components due to corrosion. No more painting or special sealing needed.

How much can you save by upgrading to a DIP System®?

Chapter 9

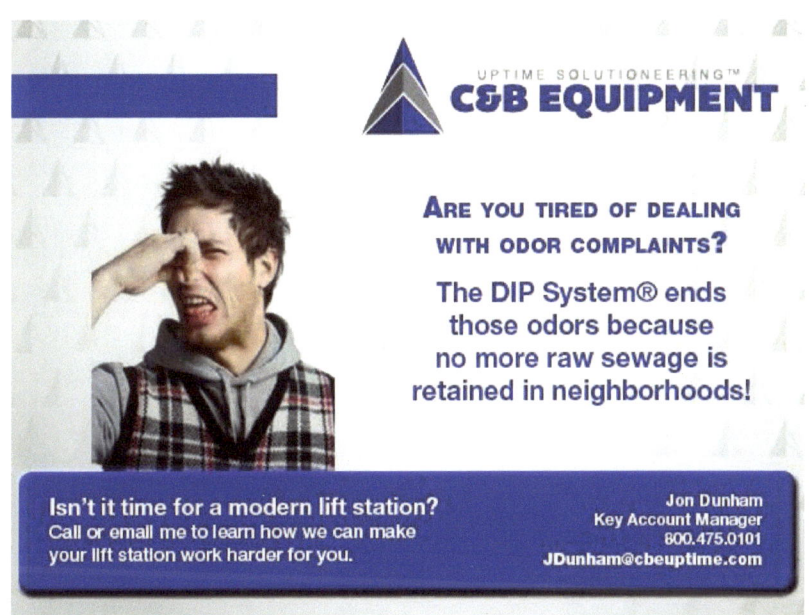

The neighborhoods will experience the added relief of no longer having the hydrogen sulfide gas odors due to the required local retention of sewage in a wet well as has been the practice through using submersible pumps for the last 60 years. The need for a wet well is also eliminated and it is converted to a clean, dry pumping system work area that is now safe for employees.

Why would you continue with an outdated wet well system?

Chapter 10

The United States has been using wet wells with submersible pumps since the early 1930's. This same "technology" is still used as part of our aging wastewater systems today. What is a wet well? It is a typically round, concrete cylinder, 8 feet or more in diameter, and about 25 feet deep and found in neighborhoods and in city lots. It will usually have two pumps (duplex) that are submerged to the bottom of the well. Sewer pipes from surrounding homes and businesses drain their raw sewage, after flushing, by gravity into the wet well. The wet well is used to collect and retain the raw sewage until a predetermined level is reached and it is then pumped down and sent on to the treatment plant. The wet well can now fill again. It is at this point that "the trouble with wipes" begins. Anything that is flushed arrives at the wet well. If the wet well inlet uses a basket or screen to filter out the "flushables" it takes manpower to pull the baskets, remove the trash, and then haul it away. Some wet wells, in some neighborhoods, need to be serviced daily. If wipes pass by the screens or baskets, they end up on the impellers of the pumps causing the pumps to fail. A crew must then be dispatched to pull, clean, and replace the pumps. Not only are the wet wells a receptacle for trash, they also produce Hydrogen Sulfide (H_2S) which is a toxic gas that causes corrosion of the equipment in the wet well. Raw sewage and wet wells produce the gas and odors that are detectable in the neighborhoods or near treatment plants.

Chapter 11

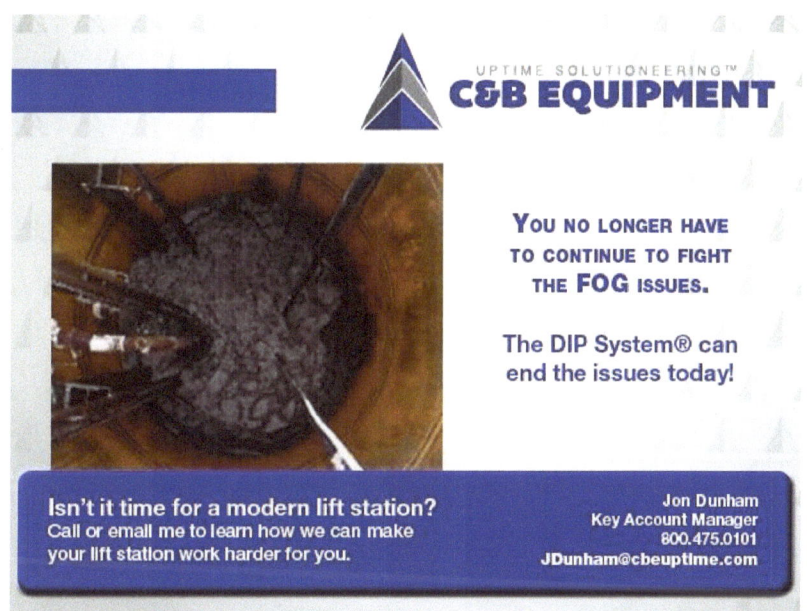

Fats, Oils, and Grease are problems when you have a wet well system. Regular expenditures for cleaning and services will continue as long as you have a wet well.

With a DIP System® these issues are no longer a problem or an added expense. Why would you want to keep paying?

Chapter 12

Anything that is flushed arrives at the wet well. If the wet well inlet uses a basket or screen to filter out the "flushables" it takes manpower to pull the baskets, remove the trash, and then haul it away. Some wet wells, in some neighborhoods, need to be serviced daily. If wipes pass by the screens or baskets, they end up on the impellers of the pumps causing the pumps to fail. A crew must then be dispatched to pull, clean, and replace the pumps.

If you have baskets or bar screen and rakes, you need to have a crew remove, load, and haul the trash on a regular basis. With a DIP System® these labor intensive and expensive services are ended. Why would you continue to spend these funds?

Chapter 13

Did you get the call?

You are ready for innovation if you have or are using wet wells to contain and retain raw sewage for pumping. As wet wells have been in use since the 1930's, it's about time to make a change. You are ready for innovation if you are still using the old submersible pumps in these wet wells. These pump types have been in use since the 1950's…Again, time to upgrade and modernize.

How will you begin to innovate, update, and modernize? Just take a quick survey. Count the number of wet wells you service. Add the hours spent to clear the trash baskets, haul the trash, and regular cleaning of the wet wells. Add the cost to pull and clear the clogged pumps and even the cleaning of the float switches. Put the dollar sign on this total. Draw a circle around it.

You can be innovative as you can now see the advantage in the rehabbing of the lift stations that are robbing you of precious time and money. Just start with one. Pick your worst station to begin with. See how much you could save if you could end the many hours of maintenance on just this one. Issues with FOG? Problems with odors? These could end when you innovate.

You should want to innovate as the actions you take will show that you are now a better manager and can actually make a difference in lowering he costs of operation, your operation.

When you innovate pumping your costs will be reduced. Your personnel will have more time for better, less "dirty" work and will be happier. With more available time additional productive projects can be accomplished, again supporting your management.

What's the risk? Since there are already 1,300 plus units operating in France alone, of a system that is patented and proven with years of use, your risk exposure is next to nothing. Your risk may be that you'll be asked questions about this innovative system that you may not have ready answers to…But, we are here to help you ease and eliminate any risk you might think you have.

I'm sure you've heard this one before…"It doesn't cost, it pays." This is actually true and can be easily proven with the actual case studies from users like you. Why would you continue to waste your money month after month, year after year? I can show you the proof of what you can save.

For you to begin is very easy. All you need to do is to call or email me to let me know when it would be convenient for us to meet. Be ready with the peak flow rates and head for the first station you want to convert and have some of the numbers of hours you are currently wasting to keep this station on line. You will be surprised when you learn just how easy this is to make you look good.

Chapter 14

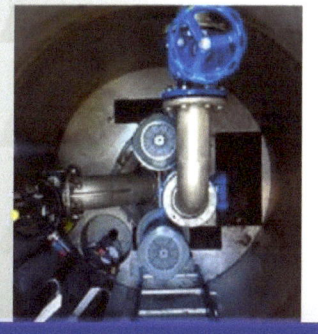

Here it is! Your former wet well that is now converted to a **DIP System®**. It's clean and dry, has no odors, is vented and well-lit, and safely accessible. Even your separate valve vault is no longer needed!

Isn't it time for a modern lift station?

Jon Dunham
Key Account Manager
800.475.0101
JDunham@cbeuptime.com

To renovate an existing wet well lift station the DIP System® adapts to any type of currently available pipework so precise positioning of input/output pipes is no longer required. The discharge head can be positioned at any angle through 360°. By eliminating the wet well, the dry tank becomes an equipment room which can be fitted with lighting, a ladder and other accessories which enable maintenance personnel to carry out their work in safety. The former wet well lift station now becomes a straightforward inspection chamber without human danger as there is no emission of dangerous gases, odors or accumulation of solid matter. The equipment is rustproof, and is thus more resistant and more durable. The absence of a collection and retention tank eliminates costly cleaning operations of those traditional units. The DIP System® enables flow or discharge capacities to be increased in a pumping station which has insufficient power, without changing the civil engineering.

Chapter 15

The DIP System® pumps can start and stop up to 150 times per hour without failing. Your submersible pumps cannot do that. Quality, design, and engineering make a difference in what you spend for pumping.

Why not use the best lift station pumping system?

Chapter 16

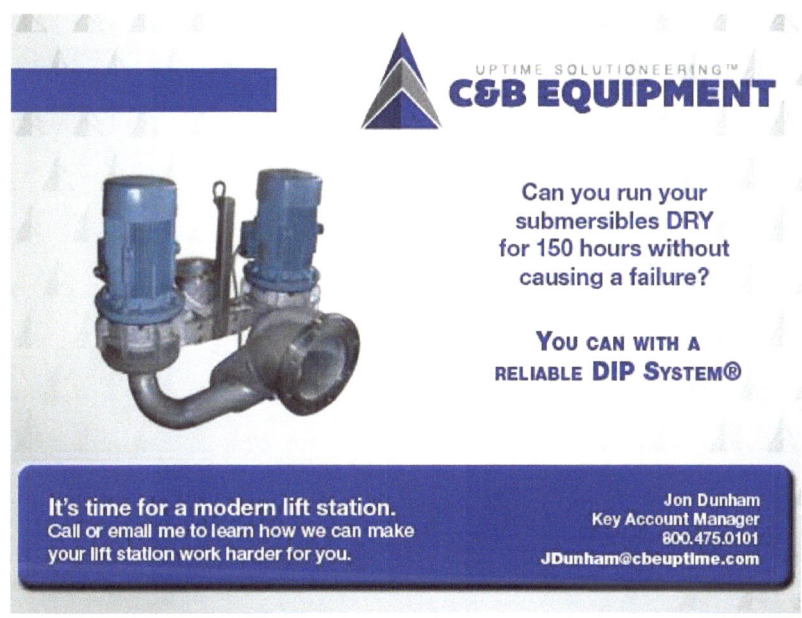

The DIP System® pumps can run dry for 150 hours with no problem. Your submersible pumps cannot do this. Why take the risk?

Chapter 17

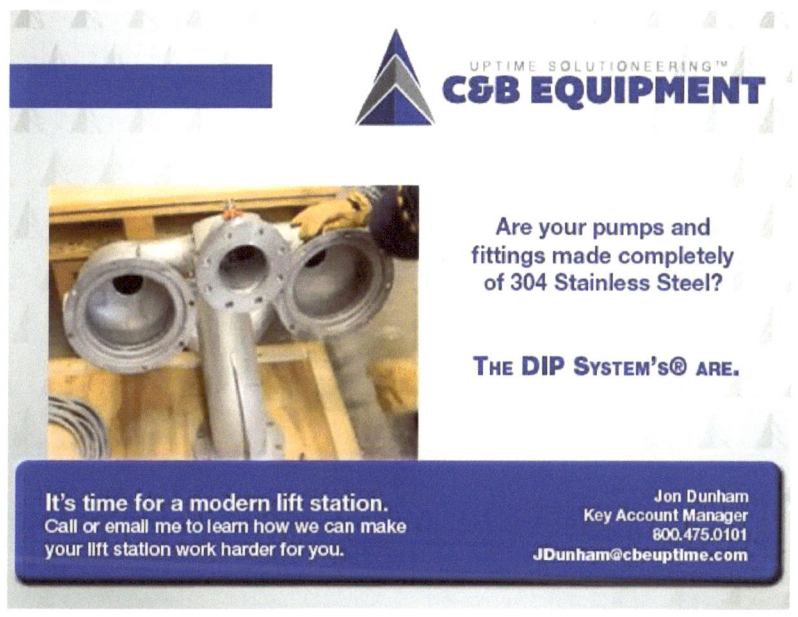

The DIP System® is made completely of stainless steel. It is made entirely from AISI 304L (316L on request) "boiler-plated" stainless steel. The bodies' suction profiles are specially designed to take advantage of the flow speed from the gravity-driven inlet.

The inlet body also includes a stone trap with inspection port and draining valve. The interior surface of the body is very smooth to improve efficiency, and has no areas where matter in suspension can collect.

The wide flow section continues through to the internal directional swing check valve. The valve box is an integral part of the body, thereby eliminating the need for collection pipework between the two pumps. The valve has three possible positions:

right or left according to which pump is operating, and central if both are operating. It has a stainless steel frame and replaceable wear plates. The discharge, from 3" to 16", depending on the model, has a standard-compliant flange and a pressure measuring socket.

A single check valve can be fitted directly to this flange.

Is your pump made of stainless steel? Why not?

Chapter 18

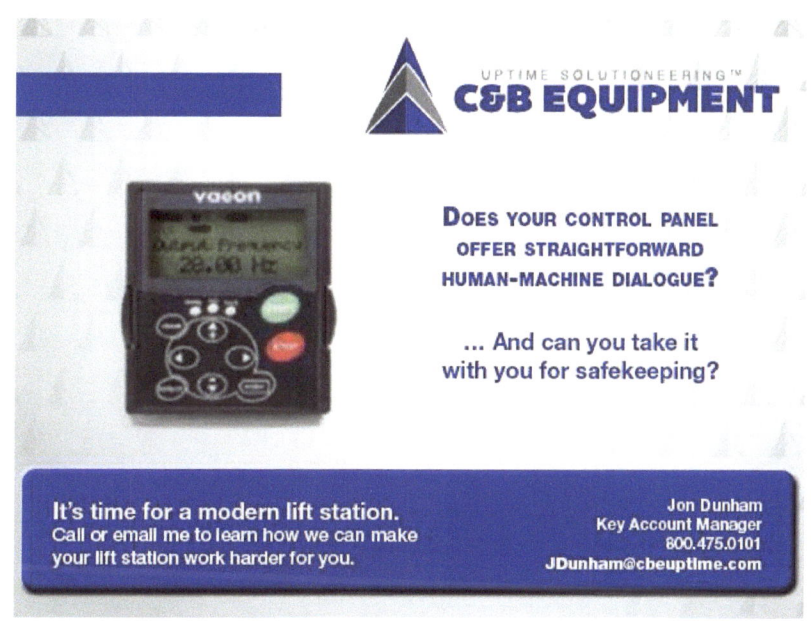

Does your control panel offer straightforward human-machine dialogue?

… And can you take it with you for safekeeping?

Wish you had a DIP System®?

Chapter 19

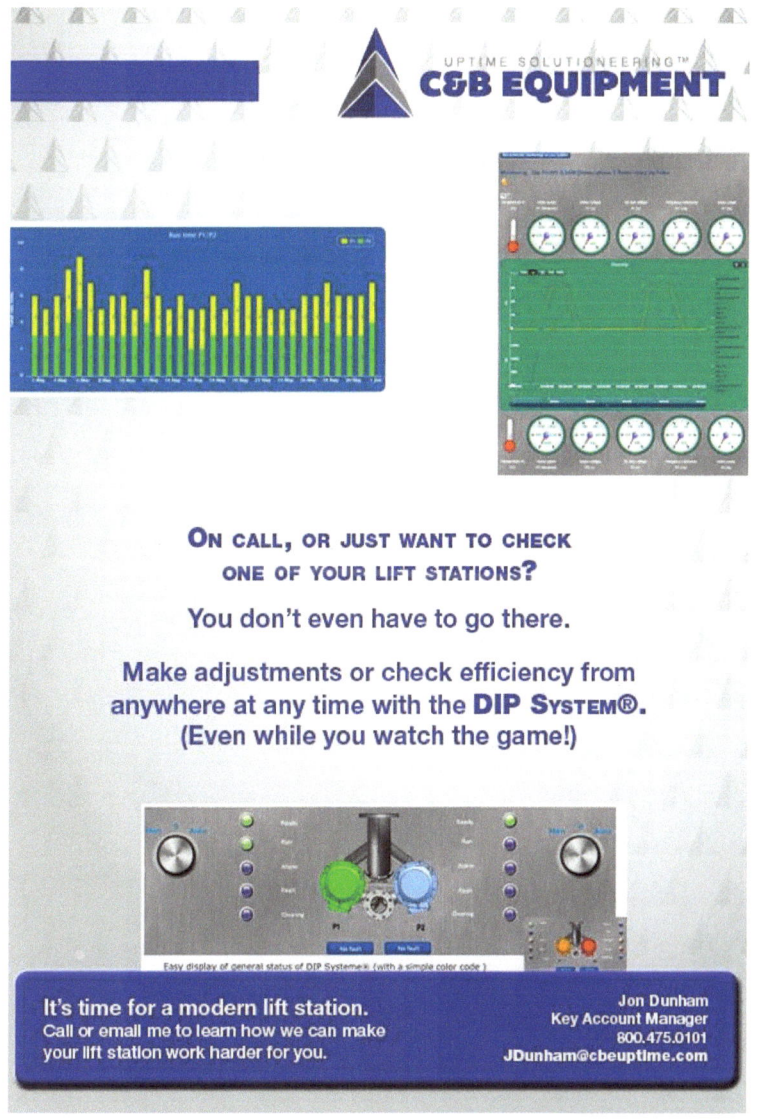

Save a trip to the lift station…Just manage remotely!

Chapter 20

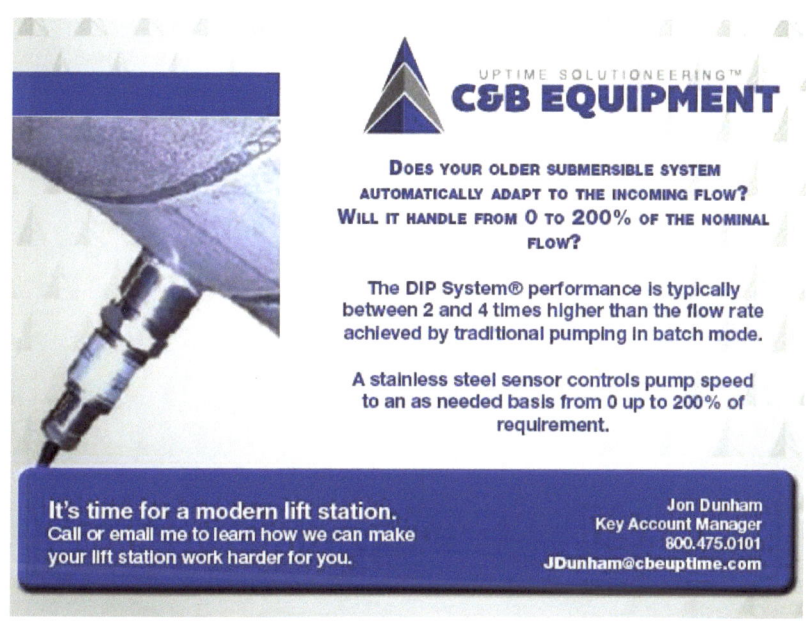

This system dispenses with the need for float switches or ultrasound measurement.

A pressure sensor located beneath the entry chamber constantly measures the height of fluid at the inlet. With its stainless steel AISI 316 flush membrane, this sensor is highly wear resistant. It is resistant to deposit build-up because it benefits from, at this position, the inlet fluid speed, which is further enhanced by the suction effect of the pump operation. Connection is via 49 ft. of IP67 protection rated cable as standard on all models. In addition to the information transmitted to the rated enclosure regarding the inlet fluid height, information from the sensor can also be sent by a transmitter for remote surveillance of the system, without the need for any accessories, through an isolated output on the control panel.

Chapter 21

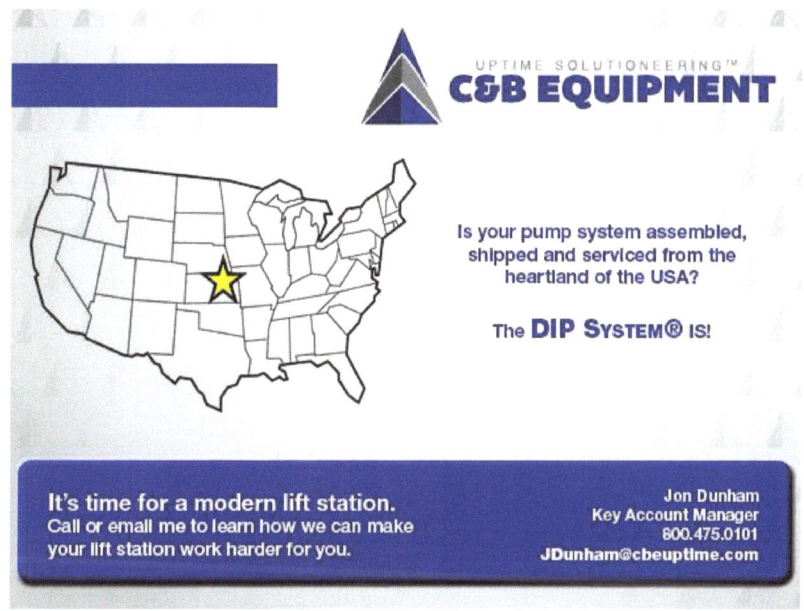

Is your pump assembled and shipped from the heart of our country? Are parts stored there too?

Why would you not want a DIP System®?

Chapter 22

Compare the DIP System® with your wet well!

Which would you rather have?

Chapter 23

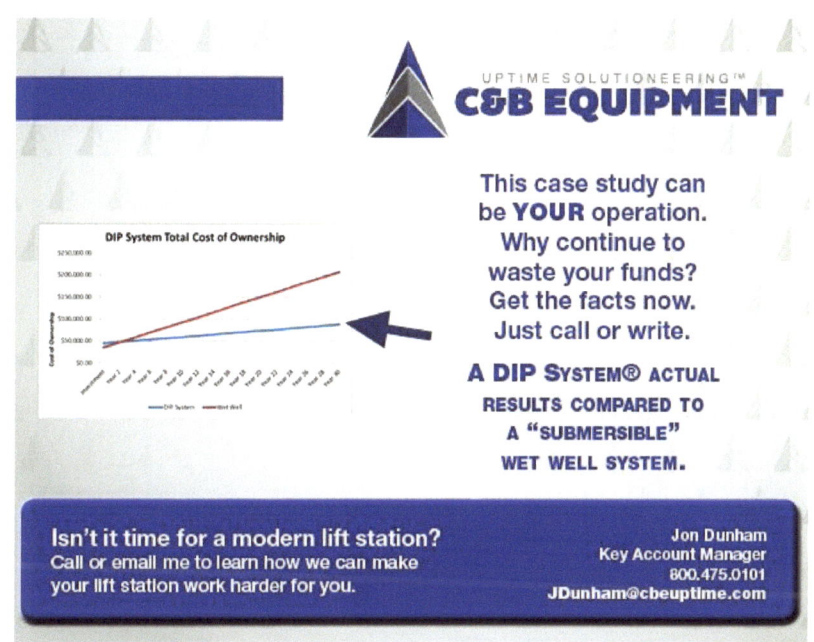

These savings can be yours. We can prove it. Why wait any longer. Contact us today to get the facts.

Chapter 24

Typical pump life cycle cost profile (Courtesy of Hydraulic Institute and pump Systems Matter)

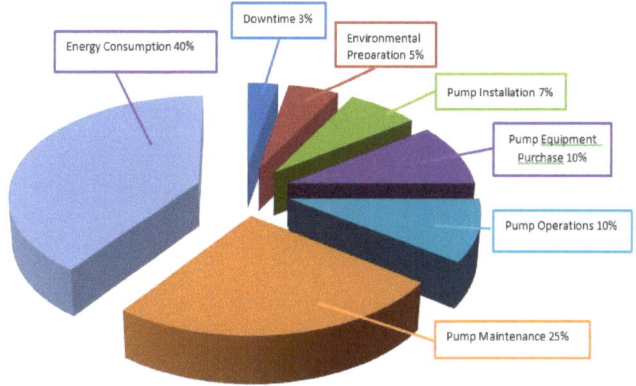

Typical pump life cycle cost profile when using DIP System® pumping lift station wastewater
What could you do with 20% more productive time per lift station?

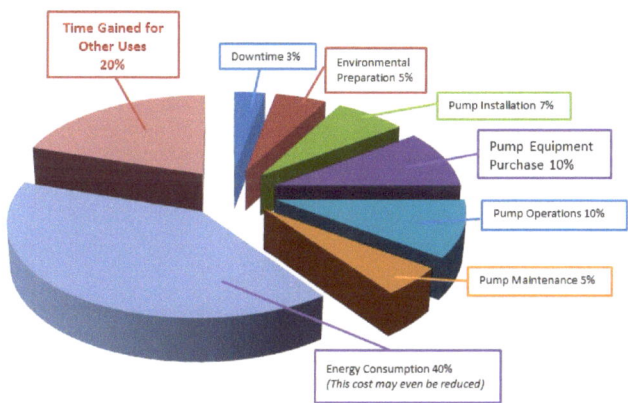

The total cost of ownership benefits shown here

Conclusion

Given the likely impending installation of many technologically advanced lift stations in the United States, municipalities will recognize that their savings of manpower costs gain greater importance.

This is done by no longer having to remove and replace clogged pumps as the DIP System® pumps are self-cleaning. Additional savings will accrue as the corrosion of lift station equipment and the station itself will cease.

The use of bar screens and rakes, trash baskets, or comminutors, all requiring regular cleaning and maintenance at the wet well is no longer needed.

The issue with "flushables" causing problems with lift station pumps will no longer need personnel attention.

The neighborhoods will experience the added relief of no longer having the hydrogen sulfide gas odors due to the required local retention of sewage in a wet well as has been the practice through using submersible pumps for the last 60 years.

The need for a wet well is also eliminated and it is converted to a clean, dry pumping system work area that is now safe for employees.

A separate valve vault is not needed and the footprint of the station is smaller.

The DIP System® lift station is remotely managed, not just monitored, as some submersible systems currently report. The management can be done via a smart phone, a tablet, or a desktop computer from anywhere. The operator has the ability to visually "see" the performance of the pumps, make any adjustments if desired, run tests, and even order printed reports to be delivered to management via remote printer from his office, the cab of his truck, or his kitchen table on a weekend. The typical daily or routine lift station "site visits" are no longer needed.

This system provides effective utilization of personnel and supports optimal usage of power in its operation. The electronic control techniques employed use soft starts and stops that eliminate water hammer shocks to valves and older force mains.

The pumps only run at the speeds needed and no longer is a wet well pump down type of application used. No large surge of wastewater will now affect the biomass of the treatment plant making the operators' job easier. The effective utilization of the voltage control techniques will maintain power quality and ensure that the pumping equipment continues to operate efficiently.

The DIP System® will contribute to maintaining an efficient lift station with high quality and reliability even under future scenarios involving varying peak loads as it is designed with 100% back-up.

About the Author

 **Jon Dunham**, Key Account Manager, C&B Equipment, Sold municipal and industrial pumps and systems as long as 40 years ago. The old submersible pumps worked at that time as there were no "flushables" produced. Energy costs were lower as were labor rates and less time was spent or needed on the submersible type systems that still are in use today. H_2S in the neighborhoods with the wet wells were problems then and still today. The DIP System® can end the old ways of doing things.

I would like to help you modernize your system. Just call or email, even if just to say "hello." Thank you.

C&B Equipment
9900 Pflumm Rd., Ste. 67
Lenexa, KS 66215
(800) 475-0101

www.cbeuptime.com

JDunham@cbeuptime.com

www.ingramcontent.com/pod-product-compliance
Lightning Source LLC
Chambersburg PA
CBHW040811200526
45159CB00022B/282